Arthi Sivan
Suresh Norman

Interface and Control of Appliances by the Analysis of
Electrooculography Signals

AF141260

Arthi Sivan
Suresh Norman

Interface and Control of Appliances by the Analysis of Electrooculography Signals

LAP LAMBERT Academic Publishing

Impressum / Imprint
Bibliografische Information der Deutschen Nationalbibliothek: Die Deutsche Nationalbibliothek verzeichnet diese Publikation in der Deutschen Nationalbibliografie; detaillierte bibliografische Daten sind im Internet über http://dnb.d-nb.de abrufbar.
Alle in diesem Buch genannten Marken und Produktnamen unterliegen warenzeichen-, marken- oder patentrechtlichem Schutz bzw. sind Warenzeichen oder eingetragene Warenzeichen der jeweiligen Inhaber. Die Wiedergabe von Marken, Produktnamen, Gebrauchsnamen, Handelsnamen, Warenbezeichnungen u.s.w. in diesem Werk berechtigt auch ohne besondere Kennzeichnung nicht zu der Annahme, dass solche Namen im Sinne der Warenzeichen- und Markenschutzgesetzgebung als frei zu betrachten wären und daher von jedermann benutzt werden dürften.

Bibliographic information published by the Deutsche Nationalbibliothek: The Deutsche Nationalbibliothek lists this publication in the Deutsche Nationalbibliografie; detailed bibliographic data are available in the Internet at http://dnb.d-nb.de.
Any brand names and product names mentioned in this book are subject to trademark, brand or patent protection and are trademarks or registered trademarks of their respective holders. The use of brand names, product names, common names, trade names, product descriptions etc. even without a particular marking in this work is in no way to be construed to mean that such names may be regarded as unrestricted in respect of trademark and brand protection legislation and could thus be used by anyone.

Coverbild / Cover image: www.ingimage.com

Verlag / Publisher:
LAP LAMBERT Academic Publishing
ist ein Imprint der / is a trademark of
OmniScriptum GmbH & Co. KG
Heinrich-Böcking-Str. 6-8, 66121 Saarbrücken, Deutschland / Germany
Email: info@lap-publishing.com

Herstellung: siehe letzte Seite /
Printed at: see last page
ISBN: 978-3-659-75216-2

Zugl. / Approved by: Chennai, Anna University, Diss. 2015

INTERFACE AND CONTROL OF APPLIANCES BY THE ANALYSIS OF ELECTOOCULOGRAPHY SIGNALS

S. V. Arthi[1], Suresh R. Norman[2]

arthikutty5@gmail.com[1], sureshrnorman@ssn.edu.in[2]

ACKNOWLEDGEMENT

To my family and friends for their immense support

TABLE OF CONTENTS

1

2

LIST OF TABLES

4

LIST OF FIGURES

LIST OF ABBREVIATIONS

ABBREVIATIONS	EXPLANATION
ADC	Analog To Digital Convertor
AREF	Analog Reference
BCI	Brain Computer Interface
DSO	Digital Storage Oscilloscope
EOG	Electrooculography
EEG	Electroencephalography
ENG	Electronystagmogram
EEPROM	Electrically Erasable Programmable Read Only Memory
HCI	Human Computer Interface
IROG	Infra Red Oculography
LED	Light Emitting Diode
PWM	Pulse Width Modulation
SRAM	Static Random Access Memory
SSC	Scleral Search Coil
USB	Universal Serial Bus
VOG	Video Oculogram

CHAPTER 1

INTRODUCTION

1.1 BACKGROUND

Communication with the outside world is important for the persons with neurological, ophthalmological disorders and paralysed patients with little motor abilities. An efficient communication and control without speech and hand movements is essential to improve the quality of life for such differently abled people. The application of using Electrooculography signals to control the HCI systems are of importance in recent decades. HCI is controlled by the bio-electric potentials produced in the body rather than normal communication pathway. There are many ways of assessing eye movements. Some of these are based on clinical observations and are qualitative.

1.2 ABSTRACT

Electrooculography (EOG) signals can be used to control human-computer interface (HCI) systems. The main objective of measuring and processing these signals is to help the users to overcome many of the limitations and inconveniences in the physical world. EOG is the methodology of tracking eye movement by placing electrodes and sensing the corneal-retinal potential (CRP), which is the resting potential between the cornea and the retina. The signals will have a voltage variation which is linearly proportional to eye displacement.

The electrodes convert the ion current obtained from the skin into electron current. The obtained bio-signal is in terms of lower voltage and hence it must be amplified, filtered and processed to remove unintended blinks, noises and other artefacts. The main purpose of this project is to analyze the extracted features such as time period and amplitude from the electrical signals corresponding to the different directions of eye movement and blinking. Hence the ultimate aim of this project corresponds to the analysis of these signals to the smart control of appliances with different types of eye movement.

1.3 EXISTING TECHNIQUES

The various techniques are Infra red oculography (IROG) where a fixed light source is directed at the eye and the amount of light reflected back to a detector varies with the eye position. It is better to measure horizontal than vertical movements because the lids cover the surface of the eye. The Video oculogram (VOG) tracks the eye movement with the camera and translates them into movements of mouse on the screen. Scleral Search coil (SSC) method has small coil of wire embedded in the modified contact lens. When a coil of wire moves in a magnetic field, the field induces a voltage in the coil. If the coil is attached to the eye, then a signal of eye position will be produced. The disadvantage is that this method is invasive.

The Electrooculography (EOG) is one of the beneficial systems that track the eye movement by the electrical signal produced by the standing potential developed between the cornea and the retina of the eye during the eye movement. This occurs due to the presence of large number of electrically active nerves in the retina compared to the front of the eye. One of the advantages of the EOG measurements over other techniques is that the field–of–view is not reduced by the glasses that are used as mounting platform for video or other sensors and illuminators.

EOG is inexpensive, easy to use, reliable, and relatively unobtrusive when compared to head-worn cameras used in video-based electrooculography. It is used because they are easier to detect. The relationship between EOG and eye movements is linear and the waveform is easy to detect. Considering the characteristics of EOG mentioned above, EOG based HCI is becoming the hotspot of bio-based HCI research in recent years. The EOG ranges from 0.05 to 3.5 mV in humans and is linearly proportional to eye displacement. An EOG-based system can be used to control a wheelchair [5], a television [6], or a keyboard [7]. Processing the EOG signals effectively helps to overcome the physical limitations and inconveniences of our daily life. This project presents the analysis of the eye gestures using EOG technique for the smart control of the appliances with the help of wired electrodes and signal conditioning circuits.

CHAPTER 2

LITERATURE SURVEY

The Electrooculography technique is traditionally used for the diagnosis of ophthalmologic, neurologic and vestibular disorders. In recent years, it has been found that EOG based Human Computer Interface applications were widely used. A detailed study of various systems implemented was reviewed as a part of this work.

The work by Ali Bulent Usakli and Serkan Gurkan, deals about the EOG based virtual keyboard [1]. The system extracts features corresponding to 4 directions and voluntary blinks for selection . By using a realized virtual keyboard, it is possible to notify in writing the needs of the patient in a relatively short time. The system software was used to transfer and classify the EOG signals in real time. When the EOG measurement starts, the horizontal and vertical EOG signals appear on the main menu. With the eye-movement virtual keyboard, needs and motion control submenus can be selected. Each cursor movement was performed step by step. One blink was enough to select the options. The ongoing eye blinks do not cause the selection. If the selection was wrong, it can be returned to the main menu. When the keyboard was selected, the virtual keyboard appears on the screen as a submenu. It was possible to use a virtual PC keyboard and a standard BCI application. However, involuntary blinks were not considered and the system used was costly.

Andreas Bulling et al. proposed a system for analysing the activity of eye movement [2]. The three different eye movement types are detected from the processed eye movement data: saccades, fixations, and blinks. The corresponding eye movement events returned by the detection algorithms are the basis for extracting different eye movement features using a sliding window. In the last stage, a hybrid

11

method selects the most relevant of these features, and uses them for classification. The ubiquity of the eyes' involvement in everything a person does means that it is challenging to annotate precisely what is being "done" at any one time. Reading was one of the easiest to capture because of the intensity of eye focus that was required and the well-defined paths that the eyes follow. A task such as Web browsing was more difficult because of the wide variety of different eye movements involved. The difficulty was to separate relevant eye movements from momentary distractions.

Andres U beda et al. designed a wireless EOG- based interface system which uses a transmitter and receiver [3]. The signal was filtered and then, the derivative was calculated to obtain the abrupt changes of the signal. Finally, using specific thresholds for each user, the direction was obtained. The accuracy of the algorithm depends on the election of the thresholds of the signal. The vertical movement was harder to detect because the linear range of movement was smaller for the vertical movement than for the horizontal. Training must be performed to obtain the proper thresholds for each channel. Blinks and diagonal movements are not considered in the algorithm. Further miniaturisation of the device could be done to make it more comfortable.

Shang-Lin Wu et al. proposed a paper about the classification method used in a wireless EOG-based HCI device for detecting eye movements in eight directions [12]. This device includes wireless EOG signal acquisition components, wet electrodes and an EOG signal classification algorithm. This algorithm was composed of three parts: signal pre-processing, feature extraction, and classification. The EOG classification algorithm is based on extracting features from the electrical signals corresponding to eight directions of eye movement (up, down, left, right, up-left, down-left, up-right, and down-right) and blinking. This system provides an effective method for identifying eye movements. However, results were affected by the

12

interference due to unexpected motions of the eye. Also, Blinking affected the performance. Additionally, it may be applied to study eye functions in real-life conditions.

Watcharin Tangsuksant et al. proposed a new EOG system for typing words in virtual keyboard [15]. Here, two experiments were used to evaluate system efficiency: one was typing speed test, and another was accuracy of eye movement direction detection. EOG signal was measured from two channels with six electrodes affixed around the eye. These 6 electrodes used were considered inconvenient for the movement of eyes. Improvements in the system such as noise robustness, accuracy and several direction detection of eye movement have to be added for increasing the performance.

CHAPTER 3

ENGINEERING ANALYSIS

3.1 INTRODUCTION

The Electooculography technique is used here for the detection of eye movement. This technique uses surface electrodes which are placed around the eyes and signals from these electrodes at five positions are taken. The chemical activity in the nerves and muscles of the body generate a variety of signals. They produce a characteristic pattern of voltage variations. This pattern is recognized for different eye movements and is used for controlling appliances. The potential difference is established in the cell and it acts similar to a tiny battery. When electrodes are placed around the eyes, they capture the resting potential between the cornea and retina thus employed for analyzing different eye movements.bio-signals in the range of milli volts are obtained which are then passed through signal conditioning circuit. The output is displayed using a digital storage oscilloscope and analysed further using a microcontroller.

3.2 BLOCK DIAGRAM

The EOG technique is used to obtain signals using electrodes at five positions around the eyes. Bio-signals are obtained in the range of 0.05 - 3.5 mV and has a useful frequency range from 0 to 16 Hz. So, it is amplified and filtered to remove noise. The output can be recorded or displayed using a digital storage oscilloscope (DSO). The block diagram of EOG signal conditioning circuit is as given in Fig 3.1

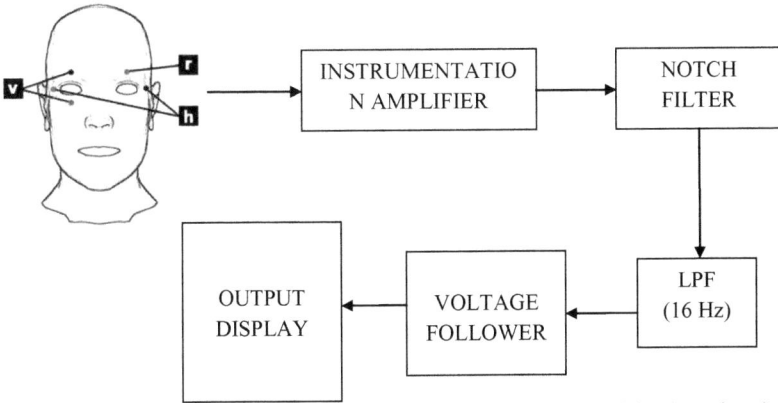

Fig 3.1 Basic Block Diagram of EOG signal conditioning circuit

The amplified and filtered signal is analyzed using the microcontroller ATMega 328. The entire block diagram of EOG signal acquisition is as shown in Fig 3.2.

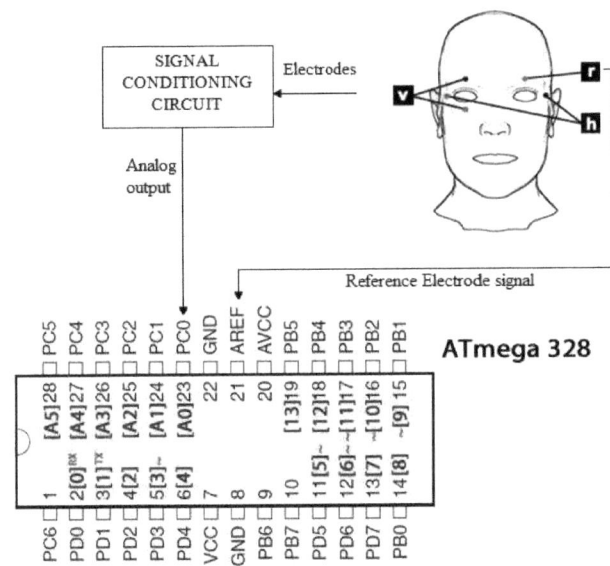

Fig 3.2 Entire Block diagram

15

Here, the output from the reference electrode is given to the analog reference input pin in order to get accurate digital data. The output from the signal conditioning circuit is given to the analog input pin. EOG signals are obtained by 4/5 electrode configuration using the EOG technique. The signal is amplified with suitable gain using an instrumentation amplifier. The high frequency and other power line noises are removed using filters. The signal is analyzed by converting the analog data into digital data by connecting it to the micro controller. Further this obtained digital data is analyzed and the pattern is recognized for different eye movements. The flowchart for the entire process is as shown in Fig 3.3.

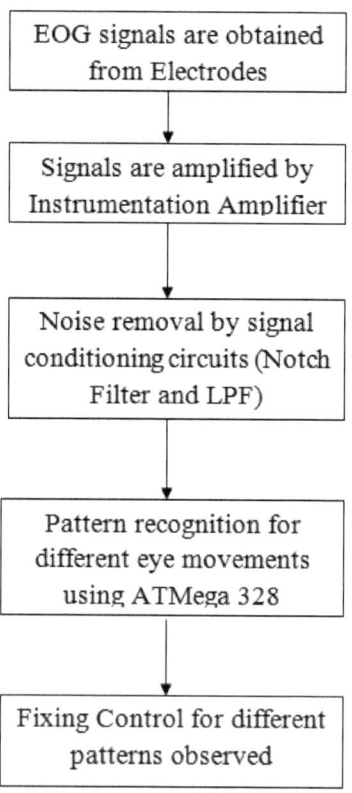

Fig 3.3 Flowchart for EOG signal acquisition and control

16

3.3 ELECTRODES

The electrodes used in medical devices transfer the ionic energy into electrical energy in the body. These currents can be amplified and have proved to be helpful in diagnosing various diseases. Medical electrodes comprises of a lead, metal and electrode conducting paste. Medical electrodes proceed with quantification of internal ionic currents and results in diagnosis of various ocular, nervous, cardiac, and muscular disorders. The device works through provision of an electrical contact between apparatus used to monitor activities and patient.

3.3.1 TYPES OF ELECTRODES

The electrodes can be segmented as reusable disc, disposable, headbands and saline based electrodes. In addition, the electrodes can also be classified based on their applications such as ECG Electrodes, Blood Gas Electrodes, EEG/EMG/ENG Electrodes, and Defibrillator Electrodes. Moreover, they can also be sub segmented as Fetal Scalp Electrodes, Electrosurgical Electrodes, TENS Electrodes, Pacemaker Electrodes, pH Electrodes, Nasopharyngeal Electrodes, and Ion-selective Electrodes. Rise in electrode applications through use of nanotechnology is augmenting the demand of electrode in various medical applications. Moreover, the treatment through this device is of minimally invasive nature and has proved to propel the growth of this market substantially.

The various types of electrodes available are wet, dry and insulating electrodes. Conductive gel is extremely useful when you use electrodes to record electrical activity on the surface of the skin. The outside layer of skin is made up of dead skin cells. These skin cells act as a good resistor, meaning that it makes detecting electrical activity difficult for EOG electrodes. Conductive Gel is usually viscous and salty. This acts as a conductive liquid medium between the skin and the electrode. In other words, the gel helps the electrode to detect the electrical activity under the skin. Artefact levels for dry and insulating electrodes were significantly higher than those for wet types. The long time stability of the skin contact is obtained using the adhesive electrodes or an electrogel. The Ag/Ag–Cl electrodes are used typically. Such electrode types are conductive, but the other types (e.g. capacitance–based) are also used. The conductive electrodes support DC signal. Here, Ag/Ag-Cl electrodes are used for good conductivity. Fig 3.4 shows (a) wet electrode and samples of dry electrodes classified as (b) stiff material (metal disc), (c) soft/flexible material (conductive polymer and foam) and (d) fabric electrode.

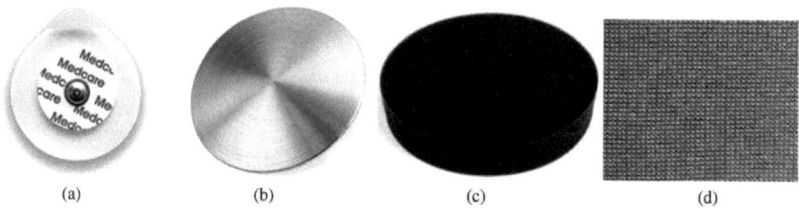

(a)   (b)   (c)   (d)

Fig 3.4 Electrode categories: (a) conventional Ag/AgCl, (b) stainless-steel disc, (c) conductive foam and (d) conductive fabric.

3.3.2 PLACING OF ELECTRODES

The properties of EOG signal varies depending on the placement of electrodes. There are many electrode configurations such as 3/4, 4/5, 7/8 configurations used for different applications. In the 3/4 configuration, there are three main electrodes, for two differential measurements: LEFT–UP and RIGHT–UP and an additional fourth reference electrode (REF). The 3/4 configuration is minimal one for the full estimation of eye orientation in both directions and blinking too. Further reduction is possible if the single orientation is sufficient for a specific application.

The 7/8 configuration is the maximal variant that allows the measurement of very precise movement including every eye separately. It is important for medical diagnosis applications. In this project, 4/5 electrode configuration is being used since it is a compromise between the other two configurations. The wires used in the 7/8 configuration especially below the eyes are a disadvantage. Generally the number of electrodes depends on the acquisition systems. The first number denotes the number of active electrodes being placed and the second number denotes the total number of electrodes including the reference electrode. The different measurement systems are as shown in Fig 3.5

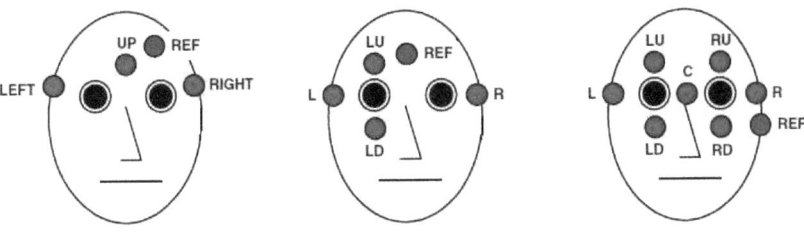

Fig 3.5 Different measurement systems: 3/4, 4/5, and 7/8

3.3.3 VERTICAL ELECTRODE PLACEMENTS

The electrode placement for detecting vertical (top-down) movement of eye is as shown in Fig 3.6. One of the electrodes is placed at the top of one eye and another electrode at the bottom of the same eye. These electrode inputs are given to Instrumentation amplifier (INA118) input pins 2 and 3. The third electrode is placed on the forehead which acts as the reference electrode. This electrode is connected to the pin 5 of the Instrumentation amplifier (INA118).

19

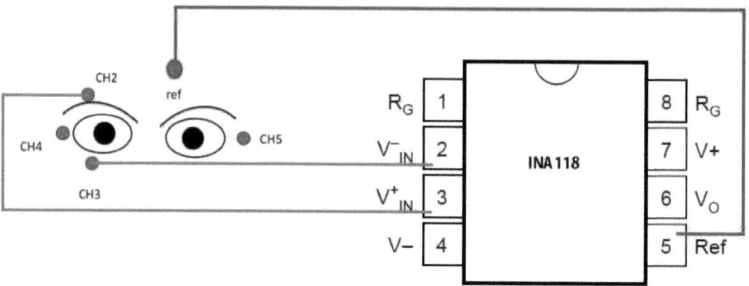

Fig 3.6 Vertical electrode placements

3.3.4 HORIZONTAL ELECTRODE PLACEMENTS

The electrode placement for detecting horizontal (left-right) movement of the eye is as shown in Fig 3.7. One of the electrodes is placed at the left of one eye and another electrode at the right side of the other eye. These electrode inputs are given to the Instrumentation amplifier (INA118) pins 2 and 3. The third electrode is placed on the forehead which acts as the reference electrode. This electrode is connected to the pin 5 of the Instrumentation amplifier (INA118).

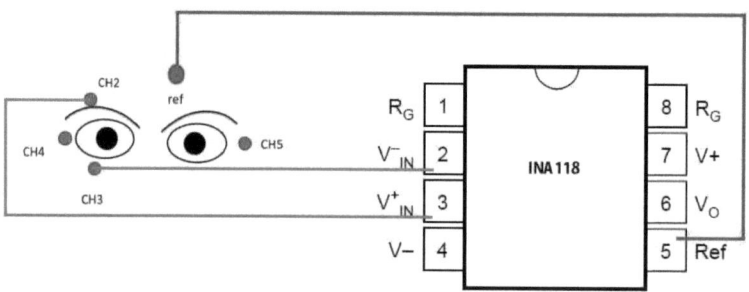

Fig 3.7 Horizontal electrode placements

20

CHAPTER 4

SIGNAL PROCESSING

4.1 INTRODUCTION

The bio-electric signals are generally of low voltages and hence the required signal conditioning must be done in order to analyse the signal. Differential amplifiers are handy tools when it comes to obtaining an EOG signal. A differential amplifier is an electronic filter that amplifies the difference between two voltages. Signal processing is an enabling technology that encompasses the algorithms and implementations of processing or transferring information contained in many different physical or abstract formats for representation, modelling, analysis and acquisition. Analog signal processing is for signals that have not been digitized, as in legacy radio, telephone, radar, and television systems. This involves linear electronic circuits such as passive and active filters as well as non-linear circuits.

4.2 INSTRUMENTATION AMPLIFIER

The INA118 is a low power, general purpose instrumentation amplifier offering required gain for the project. Its versatile 3-op amp design and small size make it ideal for a wide range of applications. Current-feedback input circuitry provides wide bandwidth even at high gain (70kHz at G = 100). A single external resistor sets any gain from 1 to 10,000. Internal input protection can withstand up to ±40V without damage. The INA118 is laser trimmed for very low offset voltage (50mV), drift (0.5mV/°C) and high common-mode rejection (110dB at G = 1000). It operates with power supplies as low as ±1.35V, and quiescent current is only

21

350mA—ideal for battery operated systems. The INA118 is available in 8-pin plastic DIP, and SO-8 surface-mount packages, specified for the –40°C to +85°C temperature range. It is one of the best options out there for biomedical systems and a powerful tool with a high gain and a high CMRR (Common mode Rejection Ratio), thus making it a perfect solution for this application. The 110dB CMRR of the amplifier at gain=1000, eliminates common signals that go in both inputs, hence removing some noise. The basic block diagram of instrumentation amplifier is as shown in Fig 4.1.

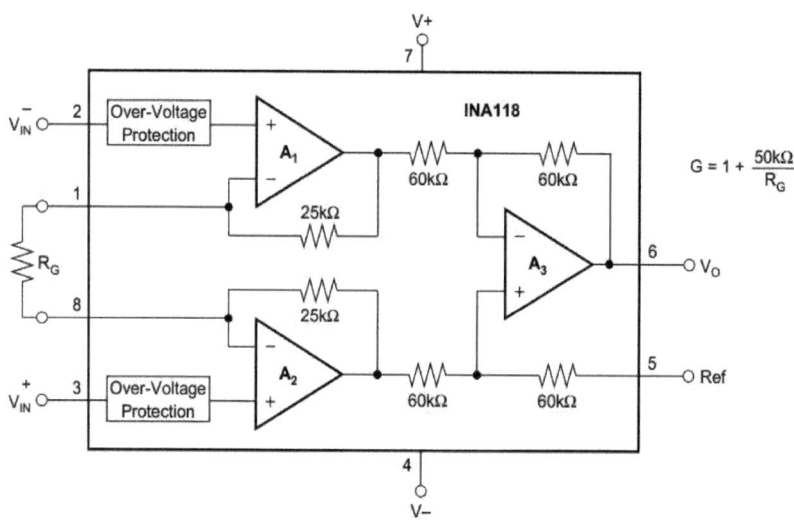

Fig 4.1 Block diagram of an instrumentation amplifier

4.3 NOTCH FILTER

Filters are electronic circuits which perform signal processing functions, specifically to remove unwanted frequency components from the signal, to enhance wanted ones, or both. It is sometimes desirable to have circuits capable of selectively filtering one frequency or range of frequencies out of a mix of different frequencies in a circuit. A circuit designed to perform this frequency selection is called a *filter circuit*, or simply a *filter*. A narrow-band band-reject filter is referred to as a notch filter and the wideband band-reject filter is referred to as band-reject filter. Twin T notch filter is one which can tune up to 100dB but it is hard to tune over a range of frequencies. The op-amp being used is LM358 which has many advantages as follows:

- Two internally compensated op amps
- Eliminates need for dual supplies
- Allows direct sensing near gnd and vout also goes to gnd
- Compatible with all forms of logic
- Power drain suitable for battery operation

The power line frequency is to be removed and hence the R and C values are chosen accordingly. The notch filter which is used is as shown in Fig 4.2. The required cut off frequency is f_0=50 Hz. Thus R_1=R_2=27 kΩ, C_1=C_2=0.1 uF, C_3=0.2 uF and R_3=56 kΩ. The formula for calculating Cut off frequency is $f_0 = 1/(2\pi RC)$.

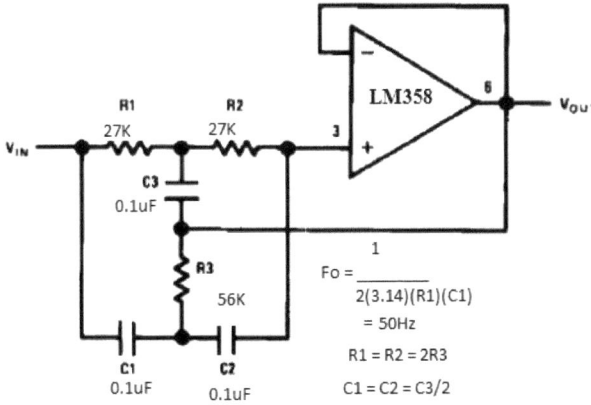

Fig 4.2 Notch filter circuit

23

The response of the notch filter will be steep and it is as shown in the frequency response curve in Fig 4.3.

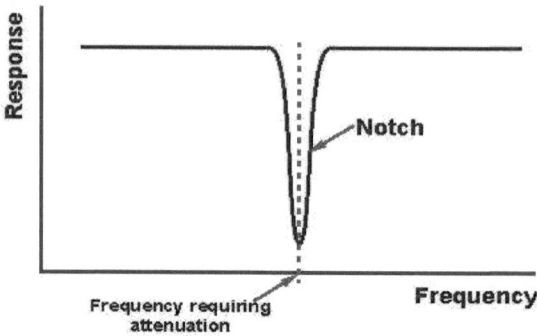

Fig 4.3 Notch filter response

4.4 LOW PASS FILTER

A low-pass filter is a device that passes low frequencies within a certain cut off and rejects (attenuates) frequencies outside that cut off frequency. The noises and other artefacts come under high frequencies and hence it must be removed. Also the EOG signals are effectively obtained only in the range of frequencies from 0 to 16 Hz. The cut off frequency is hence selected greater than but near to 16 Hz. Low pass filter response is more accurate as the order of the filter is increased. Thus a second order active low pass filter is being used here as shown in Fig 4.4. The values of resistors and capacitors are chosen accordingly to be R=100 kΩ and C=0.1 uF. The formula for calculating Cut off frequency is $f_0 = 1/ (2\pi RC)$.

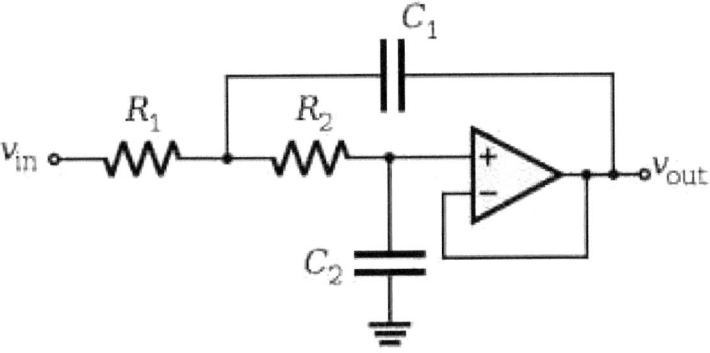

Fig 4.4 LPF circuit

4.5 VOLTAGE FOLLOWER

A voltage follower (called a buffer) is one that provides electrical impedance transformation from one circuit to another. A buffer amplifier, or simply a buffer, is an electronic amplifier that is designed to have an amplifier gain of 1. Buffers are used in Impedance matching, the benefit of which is to maximize energy transfer between circuits/systems.

There are two main kinds of buffer circuits, Voltage buffers and Current buffers. The purposes of each are to isolate the mentioned characteristic to avoid loading the input circuit or source from the output stage. The name voltage follower is given because of the characteristic of the amplifier to output a signal of the same amplitude as the input, $V_{out} = V_{in}$ (given the unity gain).It is a device which is used as an intermediate to connect a high output impedance device to a low input impedance device. It has an output voltage which is equal to the input voltage. Thus it provides isolation of output from signal source thus avoiding loading effects. A simple voltage follower using LM741 is as shown in Fig 4.5.

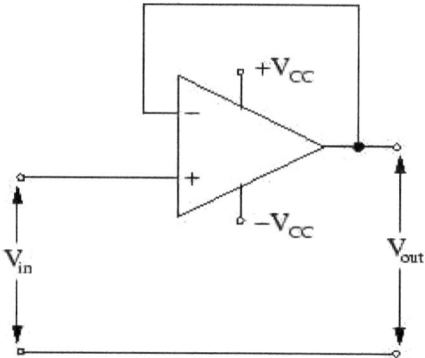

Fig 4.5 Voltage follower circuit

4.6 DC BIAS REMOVAL

There's a resting potential between the eyes. This "constant" voltage varies depending on several factors such as light, eyes' size, skin conductivity and placement of electrodes. After the amplification process, the resting potential is as well amplified, but it is something not wanted on the EOG signal.

A simple differentiator circuit can be used to remove DC *drifts* or act as a high-pass filter as shown in Fig 4.6. To remove the unwanted DC offset and to be able to read just the slope in the waveform, a small capacitor is added. The current (measured in Amperes) that passes through a capacitor is defined by:

$$I = C * dv/dt$$

When $dv/dt = 0 \rightarrow I = 0$

Therefore, $V = 0$ [$\because V = IR \rightarrow V = 0 * R \rightarrow V = 0$]

26

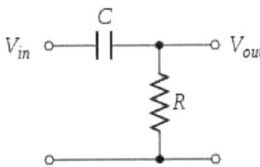

Fig 4.6 DC bias removal (HPF)

The current is defined as the capacitance times the rate of change of the voltage that passes through a capacitor, hence when the derivative of the constant potential is zero (the derivative of a constant value is = 0) the voltage after the capacitor is going to be equal to zero.

4.7 OUTPUT DISPLAY

The output can be seen by using an oscilloscope. An oscilloscope, previously called an oscillograph and informally known as a scope, CRO (for cathode-ray oscilloscope), or DSO (for the more modern digital storage oscilloscope), is a type of electronic test instrument that allows observation of constantly varying signal voltages, usually as a two-dimensional plot of one or more signals as a function of time. Non-electrical signals (such as sound or vibration) can be converted to voltages and displayed. Oscilloscopes are used to observe the change of an electrical signal over time, such that voltage and time describe a shape which is continuously graphed against a calibrated scale. The observed waveform can be analyzed for such properties as amplitude, frequency, rise time, time interval, distortion and others. Modern digital instruments may calculate and display these properties directly. Originally, calculation of these values required manually measuring the waveform against the scales built into the screen of the instrument.

A digital storage oscilloscope is an <u>oscilloscope</u> which stores and analyses the signal <u>digitally</u> rather than using <u>analogue</u> techniques. It is now the most common type of oscilloscope in use because of the advanced trigger, storage, display and measurement features which it typically provides. The principal advantage over analog storage is that the stored traces are as bright, as sharply defined, and written as quickly as non-stored traces. Traces can be stored indefinitely or written out to some external data storage device and reloaded. This allows, for example, comparison of an acquired trace from a system under test with a standard trace acquired from a known-good system. Many models can display the waveform prior to the trigger signal.

Digital oscilloscopes usually analyze waveforms and provide numerical values as well as visual displays. These values typically include <u>averages</u>, <u>maxima and minima</u>, <u>root mean square</u> and <u>frequencies</u>. They may be used to capture <u>transient signals</u> when operated in a single sweep mode, without the brightness and writing speed limitations of an <u>analog storage oscilloscope</u>. The displayed trace can be manipulated after acquisition; a portion of the display can be magnified to make fine detail more visible, or a long trace can be examined in a single display to identify areas of interest.

4.8 MICRO-CONTROLLER

The Microcontroller used here is Atmega 328 which is a cost effective version of Arduino. It has user friendly open source software and the required number of input output pins necessary to connect appliances. The pin configuration of Atmega 328 being used is as given in the Fig 4.7.

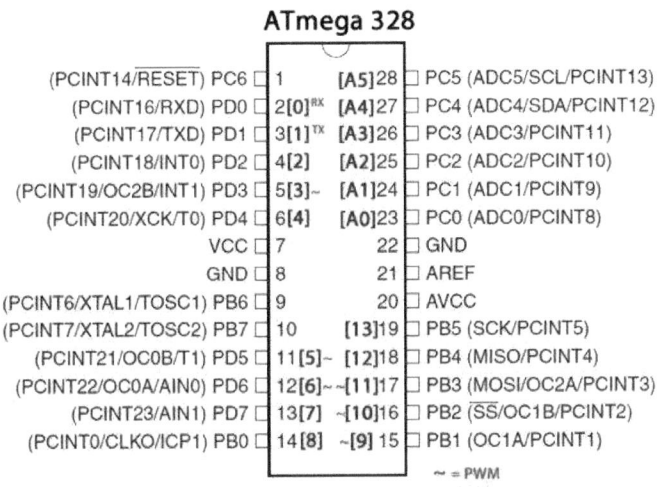

Fig 4.7 Pin configuration of Atmega 328

Atmega 328 has an in-built analog to digital convertor (ADC) which is needed for digitizing the EOG signal. Coding is done for pattern recognition and the controls are assigned for different movements. Here, ground is connected to the pin 8 or pin 22 and the analog output from the signal conditioning circuit is given to one of the six analog pins available from A0 to A5. The output EOG signal from the reference electrode is directly given to the Analog reference pin, AREF which is pin 21. Programming the Arduino board can be done by connecting it to the system using USB and so it does not need any external supply. PWM pins can be used when speed control is needed in case of motors. The digital pins can be used when LEDs, buzzers, LCDs and so on are connected.

The Arduino Uno can be powered via the USB connection or with an external power supply. The power source is selected automatically. External (non-USB) power can come either from an AC-to-DC adapter (wall-wart) or battery. The adapter can be connected by plugging a 2.1mm canter-positive plug into the board's power

29

jack. Leads from a battery can be inserted in the Gnd and Vin pin headers of the POWER connector.

The board can operate on an external supply of 6 to 20V. If supplied with less than 7V, however, the 5V pin may supply less than five volts and the board may be unstable. If using more than 12V, the voltage regulator may overheat and damage the board. The recommended range is 7 to 12 volts. Each of the 14 digital pins on the Uno can be used as an input or output, using pinMode(), digitalWrite() and digitalRead() functions. They operate at 5 volts. Each pin can provide or receive a maximum of 40 mA and has an internal pull-up resistor (disconnected by default) of 20-50 kOhms. There is a built-in LED connected to digital pin 13. When the pin is HIGH value, the LED is on, when the pin is LOW, it's off.

The Uno has 6 analog inputs, labeled A0 through A5, each of which provide 10 bits of resolution (i.e. 1024 different values). By default they measure from ground to 5 volts, though is it possible to change the upper end of their range using the AREF pin. The Arduino Uno can be programmed with the Arduino software. The Arduino Uno has a resettable polyfuse that protects your computer's USB ports from shorts and overcurrent. Although most computers provide their own internal protection, the fuse provides an extra layer of protection.

CHAPTER 5

SIGNAL ANALYSIS

5.1 INTRODUCTION

An operational amplifier needs an offset so it can work correctly, especially in this case while working with negative and positive voltages. Adding a simple offset would make the reading unstable because of the fact that the resting potential in the electrodes is not constant but depends on several environmental factors. To solve this issue, op amps need to work with a dual polarity supply, that is, positive and negative voltage with respect to virtual ground. To lower costs and complexity and also to solve the offset problem, adding a couple of 7805 voltage regulators can be done to eliminate the need for a dual power supply.

5.2 WAVEFORM ANALYSIS

The eye acts as a dipole in which the anterior pole is positive and the posterior pole is negative. Left gaze is defined as the condition when the cornea approaches the electrode near the outer canthus of the left eye, resulting in a negative-trending change in the recorded potential difference. Similarly Right gaze is defined as the condition when the cornea approaches the electrode near the inner canthus of the left eye, resulting in a positive-trending change in the recorded potential difference.

One factor to be considered is that the human eye is always active. Therefore, validation should not be performed involuntarily. To avoid such problem,

31

eye movement codification can is used. The goal of this technique is to develop control strategies based upon certain eye movements such as blinks. The normal wave when the person is looking straight is as given in the Fig 5.1. Here there is no movement of the eye and hence there is no potential developed.

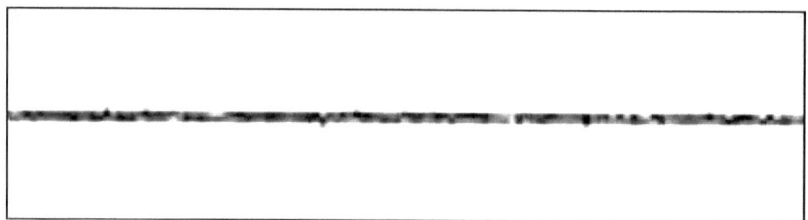

Fig 5.1 EOG wave of a person looking straight without moving the eyes

5.3 WAVEFORM ANALYSIS IN VERTICAL ELECTRODE PLACEMENT

When the electrodes are placed as given in Fig 3.4., where only the top and down movement of the eye are detected. Additionally, the blinks are dominant in this position of electrode placement. The waveform for the movement of the eye towards the top direction is inverted to that of the movement of the eye towards the bottom. The results obtained for an eye moving in the top direction is shown in the Fig 5.2 and the eye moving in the bottom is shown in the Fig 5.3.

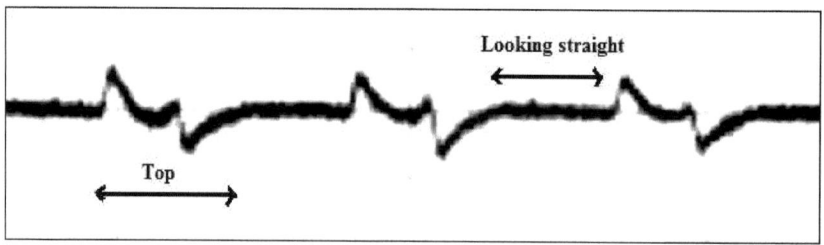

Fig 5.2 Series of top movement of the eye

32

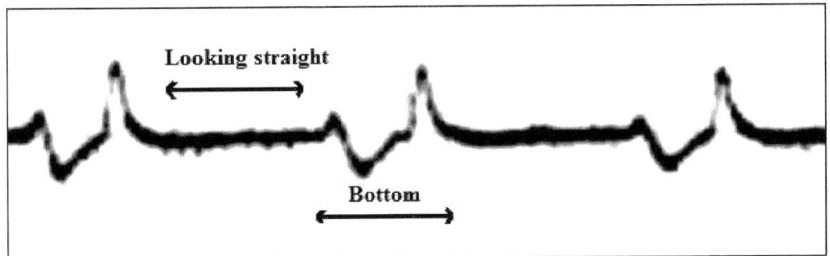

Fig 5.3 Series of bottom movement of the eye

5.4 ANALYSIS OF BLINKS

The human eye blinks are partially involuntary actions which can sometimes be made voluntarily for control purposes. In that case, the number of blinks can be counted. The continuous voluntary blinks are as shown in Fig 5.4, which have small amplitude compared with that of involuntary blinks as shown in Fig 5.5.

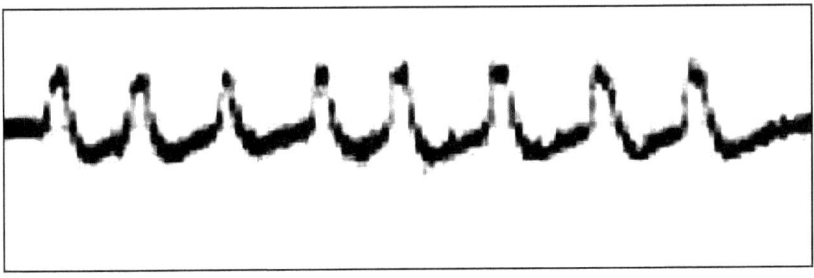

Fig 5.4 Series of Voluntary blinks at regular interval

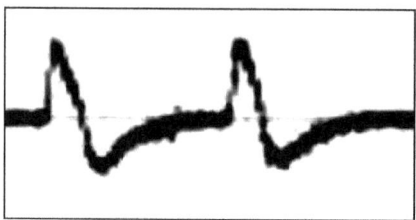

Fig 5.5 Random Involuntary blinks

5.5 WAVEFORM ANALYSIS IN HORIZONTAL ELECTRODE PLACEMENT

When the electrodes are placed as given in Fig 3.5, only the left and right movement of the eye are detected. The waveform for the movement of the eye in left direction is inverted to that of the movement of the eye in right direction. In this position of electrodes, the blinks are not much detected. The results obtained for an eye moving in the left direction is shown in the Fig 5.6 and the eye moving in the right direction is shown in the Fig 5.7.

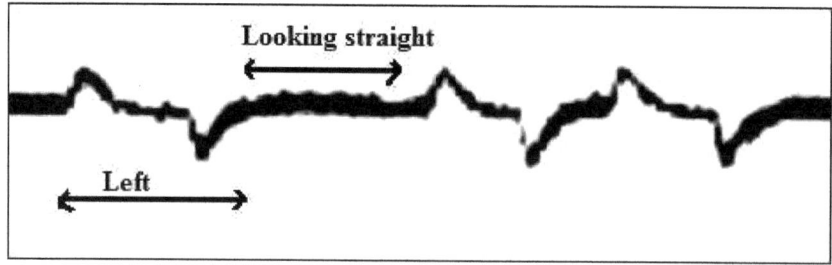

Fig 5.6 Series of left movement of the eye

34

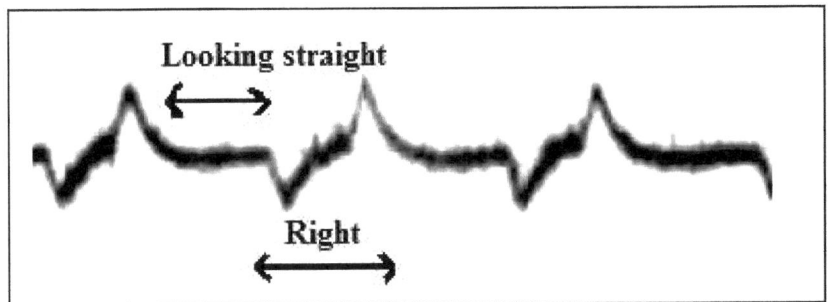

Fig 5.7 Series of waves showing the right movement of the eye

5.6 WAVEFORM ANALYSIS IN 4/5 ELECTRODE PLACEMENT

The 4/5 electrode configuration includes both the horizontal and vertical placement of electrodes. The degrees of freedom of the eye that can be detected are further increased. Thus the eye movement towards top-left, top-right, bottom-left, bottom-right can also be detected. This helps in increasing the ability to control appliances. The 7/8 electrode configuration is used rarely as there is discomfort in the eye movement due to the increased number of electrodes placed around the eyes.

CHAPTER 6

HARDWARE AND APPLIANCE CONTROL

6.1 DIGITISING THE VALUES

The analog values obtained at the analog input port of the controller are digitalized by the inbuilt ADC (Analog to Digital Controller). The resolution of the ADC is 10^{12} i.e. 0 - 1023. Thus, 4mV difference can be identified by the controller. The values obtained will be finally processed by the 16Mhz CPU. The ADC values are processed more than 600 times/second and hence it is difficult to analyze directly as the values are changing rapidly. For this purpose the resolution of the ADC is divided into ranges and each range is named. The mathematical calculations such as addition, subtraction, average and so on cannot be implemented with raw wave signal. Hence, by using greater than or less than operations we can map the values from 0-1023 in any one of the 12 ranges assigned initially.

6.2 NAMING THE RANGE OF VALUES

The ranges of values are assigned names for easy analysis (as done in ECG waveform - PQRST) and they can be used for identifying the average value over a period of time or a value at any point of time. Since average value may become nearly zero since the EOG wave is an AC signal, this method can be used to figure out the values at that point of time. This will further induce the repetition of same valued range since the processing speed of controller is 600 Hz and frequency of analog wave which we have to recognise is about 0 – 15 Hz.

36

6.3 RECORDING OF THE WAVEFORM

A change in the waveform (other than Reference) is recorded until the reference values are obtained again. Since the average of reference is above 1.2V we can set the reference of ADC as 300. Since the value of reference is also fluctuating due to noise signals from the electrode we need to add 0.4V for noise signals to define the reference from 1.2V to 1.6V. The programming is done in such a way that only if the value goes out of the range of reference value, recording of the waveform should be started temporarily until we get the reference for at least 1 second which is experimentally found as resting potential of corneal tissues. This potential may vary from person to person and also due to external conditions and noises. So, it is very important to program a range of value for identifying the reference.

6.4 FILTERING REPEATED VALUES

The repeated values obtained are filtered. This filtration is not done in the hardware but in the microcontroller programming. Since the values are repeating they fall in the same range i.e. the successive repetition of ranges are omitted. This is very important to analyze the waveform accurately. All these are done by method of string concatenation which will add all the character and join together to form a single string. Thus by comparing the obtained string to the predefined string we can actually find the characteristic of recorded waveform. The complexity of the waveform will not affect the accuracy of extracting signal from waveform since we are following the method of digital signal processing. This might not match exactly to DSP signals but point determination at particular period of time or interval is related to it.

6.5 NOISE REDUCTION

Noise reduction is done by identifying the alternatively repeating range of values. The fluctuation of values is not only present in the reference point but also during the eye movement by the resting potential and action potential. The noise present is around 0.6V during movement so we have the noise reduction vector value as 150 in the ADC. The recorded waveform as of now still has the noise signals attached to it which can be again filtered by the string filtration. The values of the waveform thus obtained for a particular interval is stored in the string.

6.6 PATTERN RECOGNITION

The process of comparing the strings will give nearly accurate waveform which is predefined. This is compared with the recorded waveform. Finally pattern recognition is done for the accepted electrooculographic waveforms. This is done by manually feeding the data into the programming loop. Thus the waveform for upward motion is detected by finding the potential difference recorded by the instrumentation amplifier which will give waveform by means of change in amplitude and frequency which is further filtered by the notch filter and also a comparator. This comparator is specially designed to compare the very small changes happening in the input of it and converts the exact ac signal in to DC wave ranging from 0 -5 V. Thus by regulating the value to the controller and fixing the Analog reference value, the controller will now detect the range and frequency without any loss.

Finally the objective of the project of finding eye movement is recognized and programming the different set of variables obtained from the recorded observation is put into working, which will differentiate between the four directions. By increasing the gain of instrumentation amplifier and making necessary changes in the filter circuits we can obtain all the respective movements, which will produce results near to video oculography but accuracy will be higher in this case.

6.7 APPLIANCE CONTROL

The left and right movements recognized in the horizontal electrode placement were tested for LED control. It was experimented by connecting an LED to the digital pin of Atmega 328 and it was programmed such that when the eyes move towards the left direction, LED was ON and similarly when the eyes move towards the right direction, LED was OFF. This can be also done with Buzzer in the place of an LED. Buzzers can be used for a patient to call someone in case of emergency. Thus connecting a device for control purely depends on the application.

The top and down movements recognized in the vertical electrode placement were also tested for LED control. An LED connected to the digital pin of Atmega 328 was programmed such that it was ON and OFF for the eyes moving in the top and down directions respectively. Further, motors can also be connected when the person is immobile. Using motors, speed control and direction control can be obtained using eye movements.

6.8 MODE CHANGE

The two modes of operation are Control mode and normal mode. The control mode operates such that all the devices connected is controlled based on the corresponding eye movement assigned in the program. The normal mode is such that no control is assigned to any of the eye movement. This mode change is facilitated when the controller encounters two continuous top movements. The experimental results were satisfactory.

CHAPTER 7

CONCLUSION AND FUTURE WORK

The technique of Electrooculography (EOG) used in the project is an inexpensive yet reliable human-computer interface that detects eye movements, a biomedical technique based on picking up signals from electrodes placed around the eyes. The EOG signals obtained in the project were amplified sufficiently using instrumentation amplifier in order to provide accurate measurements for the analysis. Further improvements in the analysis were obtained by filtering the power line noise by using a notch filter and by filtering the high frequency components by second order active low pass filter. The resulting EOG signals were obtained in the DSO and different directions of the eye movement were differentiated based on the amplitude and the time period. The results were accurate and reliable. Further noise reduction and pattern recognition were obtained by programming the microcontroller.

Thus the analysis of EOG signals and the interface lets people who cannot manipulate an object with their hands, to have more options in controlling the appliances. Further improvements can be made in placing the electrodes around the eyes such that they are more comfortable to wear. The electrodes can be fixed in a glass and can be made wearable to reduce the possibility of any errors due to the wrong positions of electrode placement or due to the wearing away of the electrode gel. The system can be made wireless by using wireless devices like Bluetooth or zigbee so that it can be used for controlling appliances with more degree of comfort.

APPENDIX 1

INSTRUMENTATION AMPLIFIER

A.1.1 FEATUES OF INA118

- Low offset voltage: 50mv max
- Low drift: 0.5mv/°c max
- Low input bias current: 5na max
- High cmr: 110db min
- Inputs protected to ±40v
- Wide supply range: ±1.35 to ±18v
- Low quiescent current: 350
- 8-Pin plastic DIP, SO-8

A.1.2 APPLICATIONS

Applications with noisy or high impedance power supplies may require decoupling capacitors close to the device pins. The output is referred to the output reference (Ref) terminal which is normally grounded. This must be a low-impedance connection to assure good common-mode rejection. A resistance of 12W in series with the Ref pin will cause a typical device to degrade to approximately 80dB CMR (G = 1). Some of the applications are:

- Bridge amplifier
- Thermocouple amplifier
- Sensor amplifier
- Medical instrumentation
- Data acquisition

The pin configuration of INA118 is as shown in Fig A.1.1

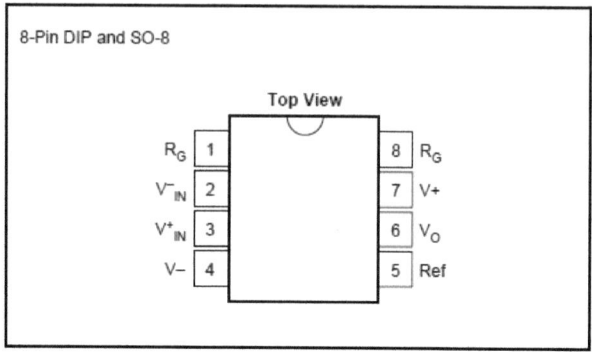

Fig A.1.1 Pin Description

A.1.3 SETTING THE GAIN

Gain of the INA118 is set by connecting a single external resistor, R_G, connected between pins 1 and 8:

$$G = 1 + \frac{50k\Omega}{R_G}$$

Commonly used gains and resistor values are given in the Table A.1.1. The stability and temperature drift of the external gain setting resistor, RG, also affects gain. RG's contribution to gain accuracy and drift can be directly inferred from the gain formula. Low resistor values required for high gain can make wiring resistance important. Sockets add to the wiring resistance which will contribute additional gain error (possibly an unstable gain error) in gains of approximately 100 or greater.

Table A.1.1 Desired gains and resistor values

DESIRED GAIN	R_G (Ω)	NEAREST 1% R_G (Ω)
1	NC	NC
2	50.00k	49.9k
5	12.50k	12.4k
10	5.556k	5.62k
20	2.632k	2.61k
50	1.02k	1.02k
100	505.1	511
200	251.3	249
500	100.2	100
1000	50.05	49.9
2000	25.01	24.9
5000	10.00	10
10000	5.001	4.99

NC: No Connection.

A.1.4 INPUT COMMON-MODE RANGE

The linear input voltage range of the input circuitry of the INA118 is from approximately 0.6V below the positive supply voltage to 1V above the negative supply. As a differential input voltage causes the output voltage to increase, however, the linear input range will be limited by the output voltage swing of amplifiers A1 and A2. Thus, the linear common-mode input range is related to the output voltage of the complete amplifier. This behaviour also depends on supply voltage. Input-overload can produce an output voltage that appears normal. For example, if an input overload condition drives both input amplifiers to their positive output swing limit, the difference voltage measured by the output amplifier will be near zero.

The INA118 can be operated on power supplies as low as ±1.35V. Performance of the INA118 remains excellent with power supplies ranging from ±1.35V to ±18V. Most parameters vary only slightly throughout this supply voltage range. Operation at very low supply voltage requires careful attention to assure that

43

the input voltages remain within their linear range. Voltage swing requirements of internal nodes limit the input common mode range with low power supply voltage.

A.1.5 DYNAMIC PERFORMANCE

The INA118 achieves wide bandwidth, despite its low quiescent current and even at high gain. This is due to the current-feedback topology of the INA118. Settling time also remains excellent at high gain. The INA118 exhibits approximately 3dB peaking at 500kHz in unity gain. This is a result of its current-feedback topology and is not an indication of instability. Unlike an op amp with poor phase margin, the rise in response is a predictable +6dB/octave due to a response zero. A simple pole at 300kHz or lower will produce a flat passband unity gain response.

A.1.6 NOISE PERFORMANCE

The INA118 provides very low noise in most applications. For differential source impedances less than 1kW, the INA103 may provide lower noise. For source impedances greater than 50kW, the INA111 FET-Input Instrumentation Amplifier may provide lower noise. Low frequency noise of the INA118 is approximately 0.28mVp-p measured from 0.1 to 10Hz (G³100). This provides dramatically improved noise when compared to state of- the-art chopper-stabilized amplifiers.

44

APPENDIX 2

OPERATIONAL AMPLIFIER

A.2.1 FEATUES OF LM358

- Available in 8-Bump DSBGA Chip- Sized Package
- Internally frequency compensated for Unity gain
- Large DC Voltage Gain: 100 dB
- Wide Bandwidth (Unity Gain): 1 MHz
- Single Supply: 3V to 32V
- Dual Supplies: ± 1.5V to ± 16V
- Very Low Supply Current Drain (500 µA)
- Low Input Offset Voltage: 2 mV
- Input common-mode voltage range includes ground
- Differential input voltage range equal to power supply voltage
- Large output voltage swing

A.2.2 UNIQUE CHARACTERISTICS

Some characteristic features unique to LM358 are as given below:

- In the linear mode the input common-mode voltage range includes ground and the output voltage can also swing to ground, even though operated from only a single power supply voltage.
- The unity gain cross frequency is temperature compensated.
- The input bias current is also temperature compensated.

A.2.3 ADVANTAGES

- The integrated circuit has two internally compensated op-amps as shown in the pin configuration in Fig A.2.1

- Eliminates need for dual supplies

- Allows direct sensing near GND and Vout also goes to GND

- Compatible with all forms of logic

- Power drain suitable for battery operation

The pin configuration of the op-amp IC being used is shown in Fig A.2.1.

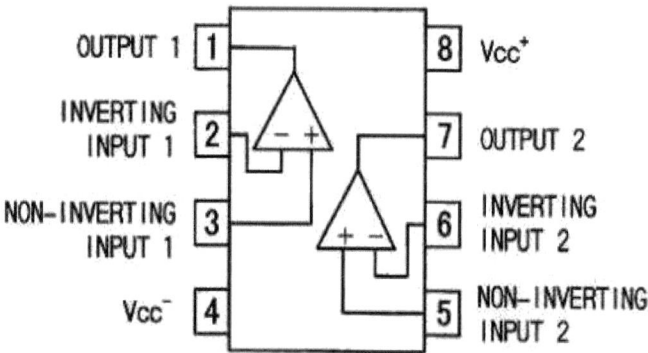

Fig A.2.1 Pin configuration of LM358

The LM358 series consists of two independent, high gain, internally frequency compensated operational amplifiers which were designed specifically to operate from a single power supply over a wide range of voltages. Operation from split power supplies is also possible and the low power supply current drain is independent of the magnitude of the power supply voltage.

Application areas include transducer amplifiers, dc gain blocks and all the conventional op amp circuits which now can be more easily implemented in single power supply systems. For example, the LM158 series can be directly operated off of the standard +5V power supply voltage which is used in digital systems and will easily provide the required interface electronics without requiring the additional ±15V power supplies. The LM358 and LM2904 are available in a chip sized package (8-Bump DSBGA) using TI's DSBGA package technology.

A.2.4 APPLICATION HINTS

The LM158 series are op amps which operate with only a single power supply voltage, have true-differential inputs, and remain in the linear mode with an input common-mode voltage of 0 VDC. These amplifiers operate over a wide range of power supply voltage with little change in performance characteristics. At 25°C amplifier operation is possible down to a minimum supply voltage of 2.3 VDC.

Precautions should be taken to insure that the power supply for the integrated circuit never becomes reversed in polarity or that the unit is not inadvertently installed backwards in a test socket as an unlimited current surge through the resulting forward diode within the IC could cause fusing of the internal conductors and result in a destroyed unit. Large differential input voltages can be easily accommodated and, as input differential voltage protection diodes are not needed, no large input currents result from large differential input voltages. The differential input voltage may be larger than V+ without damaging the device.

For ac applications, where the load is capacitively coupled to the output of the amplifier, a resistor should be used, from the output of the amplifier to ground to increase the class A bias current and prevent crossover distortion. Where the load is directly coupled, as in dc applications, there is no crossover distortion. Capacitive loads which are applied directly to the output of the amplifier reduce the loop stability margin. Values of 50 pF can be accommodated using the worst-case non-inverting unity gain connection. Large closed loop gains or resistive isolation should be used if larger load capacitance must be driven by the amplifier.

The bias network of the LM358 establishes a drain current which is independent of the magnitude of the power supply voltage over the range of 3 VDC to 30 VDC. Output short circuits either to ground or to the positive power supply should be of short time duration. Units can be destroyed, not as a result of the short circuit current causing metal fusing, but rather due to the large increase in IC chip dissipation which will cause eventual failure due to excessive function temperatures. Putting direct short-circuits on more than one amplifier at a time will increase the total IC power dissipation to destructive levels, if not properly protected with external dissipation limiting resistors in series with the output leads of the amplifiers. The larger value of output source current which is available at 25°C provides a larger output current capability at elevated temperatures than a standard IC op amp.

A.2.5 ELECTRICAL CHARACTERICTICS

The various parameters and the conditions for LM358 are listed in the Table:A.2.1

$V^+ = +5.0V$, unless otherwise stated.

Table: A.2.1 Electrical characteristics of LM358

Parameter	Conditions	LM358		
		Min	Typ	Max
Input Offset Voltage	See[1] , T_A = 25°C		2	7
Input Bias Current	$I_{IN(+)}$ or $I_{IN(-)}$, T_A = 25°C, V_{CM} = 0V, See[2]		45	250
Input Offset Current	$I_{IN(+)} - I_{IN(-)}$, V_{CM} = 0V, T_A = 25°C		5	50
Input Common-Mode Voltage Range	V^+ = 30V, See[3] (LM2904, V^+ = 26V), T_A = 25°C	0		V^+−1.5
Supply Current	Over Full Temperature Range			
	R_L = ∞ on All Op Amps			
	V^+ = 30V (LM2904 V^+ = 26V)		1	2
	V^+ = 5V		0.5	1.2

APPENDIX 3

MICROCONTROLLER

A.3.1 FEATUES OF ATMEGA 328

- High Performance, Low Power Atmel AVR 8-Bit Microcontroller
- Advanced RISC Architecture
  - ➢ 131 Powerful Instructions – Most Single Clock Cycle Execution
  - ➢ 32 x 8 General Purpose Working Registers
  - ➢ Fully Static Operation
  - ➢ Up to 20 MIPS Throughput at 20MHz
  - ➢ On-chip 2-cycle Multiplier
- High Endurance Non-volatile Memory Segments
  - ➢ 4/8/16/32KBytes of In-System Self-Programmable Flash program memory
  - ➢ 256/512/512/1KBytes EEPROM
  - ➢ 512/1K/1K/2KBytes Internal SRAM
  - ➢ Write/Erase Cycles: 10,000 Flash/100,000 EEPROM
  - ➢ Optional Boot Code Section with Independent Lock Bits
  - ➢ Programming Lock for Software Security
- Atmel® QTouch® library support
  - ➢ Capacitive touch buttons, sliders and wheels
  - ➢ QTouch and QMatrix® acquisition
  - ➢ Up to 64 sense channels
- Peripheral Features

- Two 8-bit Timer/Counters with Separate Prescaler and Compare Mode
- One 16-bit Timer/Counter with Separate Prescaler, Compare Mode, and Capture Mode
- Real Time Counter with Separate Oscillator
- Six PWM Channels
- 8-channel 10-bit ADC in TQFP and QFN/MLF package
- 6-channel 10-bit ADC in PDIP Package
- Programmable Serial USART

A.3.2 BLOCK DIAGRAM

The ATmega328 is a low-power CMOS 8-bit microcontroller based on the AVR enhanced RISC architecture. By executing powerful instructions in a single clock cycle, the ATmega328 achieves throughputs approaching 1 MIPS per MHz allowing the system designed to optimize power consumption versus processing speed. The block diagram of Atmega 328 is as shown in Fig A.3.1

The AVR core combines a rich instruction set with 32 general purpose working registers. All the 32 registers are directly connected to the Arithmetic Logic Unit (ALU), allowing two independent registers to be accessed in one single instruction executed in one clock cycle. The resulting architecture is more code efficient while achieving throughputs up to ten times faster than conventional CISC microcontrollers. The ATmega 328 provides the following features: 8Kbytes of In-System Programmable Flash with Read-While-Write capabilities, 1Kbytes EEPROM, 2Kbytes SRAM, 23 general purpose I/O lines, 32 general purpose working registers, three flexible Timer/Counters with compare modes, internal and external interrupts, a serial programmable USART, a byte-oriented 2-wire Serial Interface, an SPI serial port, a 6-channel 10-bit ADC (8 channels in TQFP and

QFN/MLF packages), a programmable Watchdog Timer with internal Oscillator, and five software selectable power saving modes

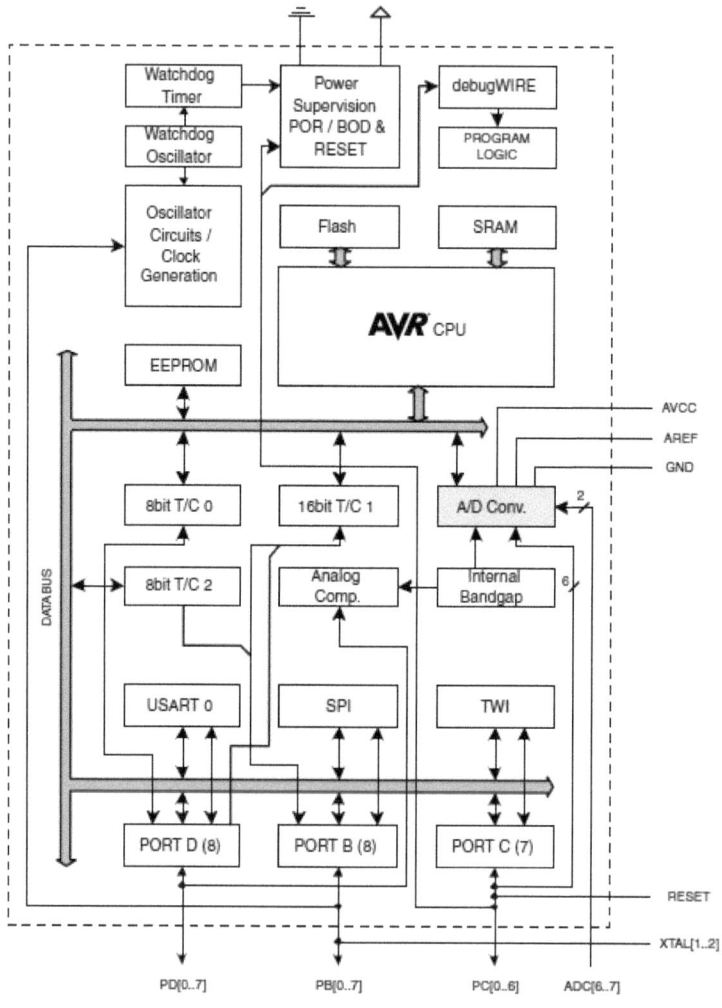

Fig A.3.1 Block diagram of Atmega 328

. The Idle mode stops the CPU while allowing the SRAM, Timer/Counters, USART, 2-wire Serial Interface, SPI port, and interrupt system to continue functioning. The Power-down mode saves the register contents but freezes the Oscillator, disabling all other chip functions until the next interrupt or hardware reset. In Power-save mode, the asynchronous timer continues to run, allowing the user to maintain a timer base while the rest of the device is sleeping. The ADC Noise Reduction mode stops the CPU and all I/O modules except asynchronous timer and ADC, to minimize switching noise during ADC conversions. In Standby mode, the crystal/resonator Oscillator is running while the rest of the device is sleeping. This allows very fast start-up combined with low power consumption.

Atmel offers the QTouch library for embedding capacitive touch buttons, sliders and wheels functionality into AVR microcontrollers. The patented charge-transfer signal acquisition offers robust sensing and includes fully debounced reporting of touch keys and includes Adjacent Key Suppression (AKS) technology for unambiguous detection of key events. The easy-to-use QTouch Suite tool chain allows you to explore, develop and debug your own touch applications.

The device is manufactured using Atmel's high density non-volatile memory technology. The On-chip ISP Flash allows the program memory to be reprogrammed In-System through an SPI serial interface, by a conventional non-volatile memory programmer, or by an On-chip Boot program running on the AVR core. The Boot program can use any interface to download the application program in the Application Flash memory.

Software in the Boot Flash section will continue to run while the Application Flash section is updated, providing true Read- While-Write operation. By combining an 8-bit RISC CPU with In-System Self-Programmable Flash on a monolithic chip, the Atmel ATmega 328 is a powerful microcontroller that provides a highly flexible and cost effective solution to many embedded control applications. The ATmega 328 AVR is supported with a full suite of program and system development tools including: C Compilers, Macro Assemblers, Program Debugger/Simulators, In-Circuit Emulators, and Evaluation kits. Reliability

Qualification results show that the projected data retention failure rate is much less than 1 PPM over 20 years at 85°C or 100 years at 25°C.

A.3.7 AVR MEMORIES

The AVR architecture has two main memory spaces, the Data Memory and the Program Memory space. In addition, the ATmega 328 features an EEPROM Memory for data storage. All three memory spaces are linear and regular.

- **In-System Reprogrammable Flash Program Memory**

The ATmega 328 contains 32Kbytes On-chip In-System Reprogrammable Flash memory for program storage. Since all AVR instructions are 16 or 32 bits wide, the Flash is organized as 16K x 16. For software security, the Flash Program memory space is divided into two sections, Boot Loader Section and Application Program Section. The Flash memory has an endurance of at least 10,000 write/erase cycles. The ATmega 328 Program Counter (PC) is 14 bits wide, thus addressing the 16K program memory locations. The program memory mapping is as shown in Fig A.3.2

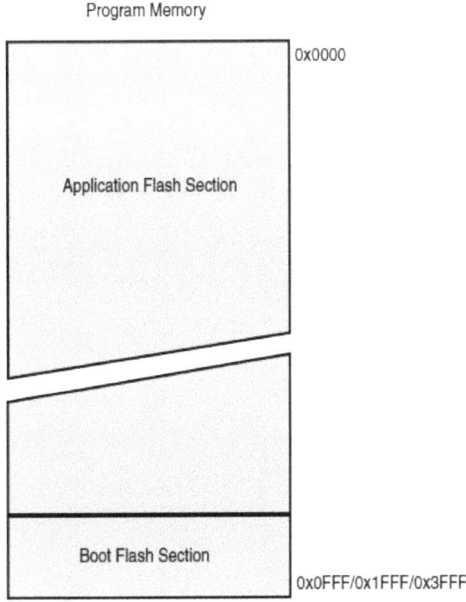

Program Memory

0x0000

Application Flash Section

Boot Flash Section

0x0FFF/0x1FFF/0x3FFF

Fig A.3.2 Program Memory Map of ATmega328

• SRAM Data Memory

The ATmega 328 is a complex microcontroller with more peripheral units than can be supported within the 64 locations reserved in the Opcode for the IN and OUT instructions. For the Extended I/O space from 0x60 - 0xFF in SRAM, only the STD and LDD instructions can be used. The lower 2303 data memory locations address both the Register File, the I/O memory, Extended I/O memory, and the internal data SRAM. The first 32 locations address the Register File, the next 64 location the standard I/O memory, then 160 locations of Extended I/O memory, and the next 2048 locations address the internal data SRAM. The five different addressing modes for the data memory cover: Direct, Indirect with Displacement, Indirect, Indirect with Pre-decrement, and Indirect with Post-increment. In the Register File,

55

registers R26 to R31 feature the indirect addressing pointer registers. The direct addressing reaches the entire data space. The Indirect with Displacement mode reaches 63 address locations from the base address given by the Y- or Z register. When using register indirect addressing modes with automatic pre-decrement and post-increment, the address registers X, Y, and Z are decremented or incremented. The 32 general purpose working registers, 64 I/O Registers, 160 Extended I/O Registers, and the 2048 bytes of internal data SRAM in the ATmega 328 is accessible through all these addressing. The data memory map is as shown in Fig A.3.3

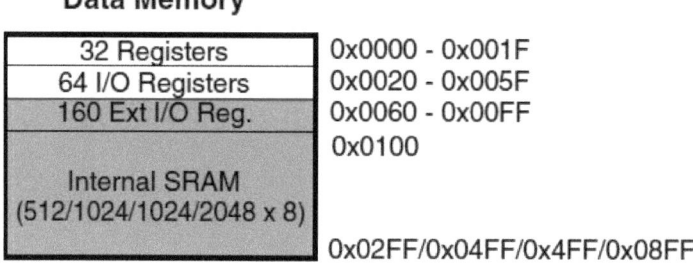

Fig A.3.3 Data Memory Map

- **EEPROM Data Memory**

The ATmega 328 contains 1Kbytes of data EEPROM memory. It is organized as a separate data space, in which single bytes can be read and written. The EEPROM has an endurance of at least 100,000 write/erase cycles. The access between the EEPROM and the CPU is described in the following, specifying the EEPROM Address Registers, the EEPROM Data Register, and the EEPROM Control Register.

• I/O Memory

The ATmega 328 I/Os and peripherals are placed in the I/O space. All I/O locations may be accessed by the LDD and STD instructions, transferring data between the 32 general purpose working registers and the I/O space. I/O Registers within the address range 0x00 - 0x1F are directly bit accessible using the SBI and CBI instructions. In these registers, the value of single bits can be checked by using the SBIS and SBIC instructions. Refer to the instruction set section for more details. When using the I/O specific commands IN and OUT, the I/O addresses 0x00 - 0x3F must be used. When addressing I/O Registers as data space using LD and ST instructions, 0x20 must be added to these addresses.

The ATmega 328 is a complex microcontroller with more peripheral units than can be supported within the 64 location reserved in Opcode for the IN and OUT instructions. For the Extended I/O space from 0x60 - 0xFF in SRAM, only the STD and LDD instructions can be used. For compatibility with future devices, reserved bits should be written to zero if accessed. Reserved I/O memory addresses should never be written. Some of the Status Flags are cleared by writing a logical one to them. Note that, unlike most other AVRs, the CBI and SBI instructions will only operate on the specified bit, and can therefore be used on registers containing such Status Flags. The CBI and SBI instructions work with registers 0x00 to 0x1F only.

APPENDIX 4

HARDWARE CONNECTIONS

A.4.1 EOG CIRCUIT

The hardware connection of the circuit for the EOG signal acquisition is as given in Fig.A.4.1. It consists of two 7805 IC to eliminate need for dual power supply. The signal conditioning circuit is implemented by the instrumentation amplifier INA118 and the filters using LM358.

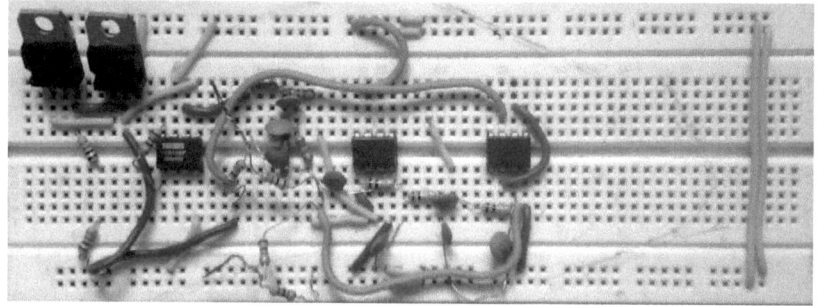

Fig: A.4.1 Hardware connection of the circuit

A.4.2 MEDICAL ELECTRODES

The medical electrodes which are used for sensing EOG signals are given in Fig.A.4.2. Here, Ag/AgCl electrodes are used. It has a lock type of pin to connect to the probe which helps in effective sensing of the signal without external noise.

58

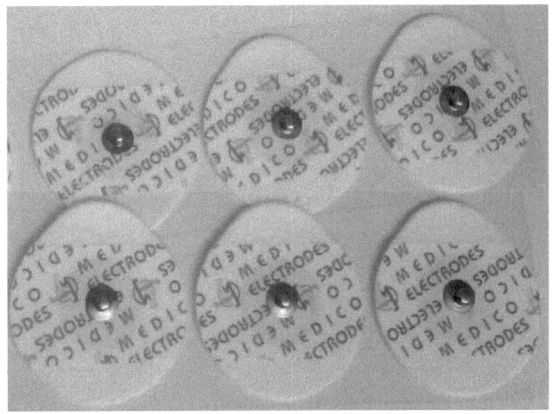

Fig: A.4.2 Medical electrodes for EOG

A.4.3 EOG PROBES

The signals from the EOG electrodes are collected through probes as shown in Fig.A.4.3. One end of the probe is locked with the electrode pin to obtain the signal and the other end is connected to the input of the instrumentation amplifier using connecting wires.

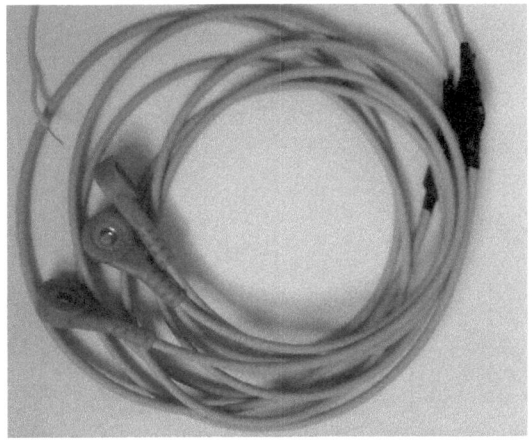

Fig: A.4.3 Probes

REFERENCES

[1] Ali Bulent Usakli and Serkan Gurkan(2010), "Design of a Novel Efficient Human–Computer Interface: An Electrooculagram Based Virtual Keyboard" , IEEE transactions on instrumentation and measurement, vol. 59, no. 8, pp. 2099 – 2108.

[2] Andreas Bulling, Jamie A. Ward, Hans Gellersen, and Gerhard Troster(2011), "Eye Movement Analysis for Activity Recognition Using Electrooculography," IEEE TRANSACTIONS on Pattern analysis and Machine Intelligence, vol. 33, no.4.

[3] Andres U beda, Eduardo Ianez, and Jose M. Azorin(2011), "Wireless and Portable EOG Based Interface for Assisting Disabled People," IEEE/ASME Transactions on Mechatronics, vol. 16, no. 5.

[4] Barae. R, Boquete. L and Mazo. M (2002), "System for assisted mobility using eye movements based on electrooculography," *IEEE Trans. Neural Syst. Rehabil. Eng.*, vol. 10, no. 4, pp. 209–218.

[5] Barea. R, Boquete. L, Mazo. M, and Lpez. E.(2002), "Wheelchair guidance strategies using EOG," *J. Intell. Robot. Syst.*, vol. 34, pp. 279–299.

[6] Deng. L. Y, Hsu. C. L, Lin. T. C, Tuan. J. S and Chang. S. M.(2010), "EOG based human–computer interface system development," *Expert Syst. Appl.*, vol. 37, pp. 3337–3343.

[7] Dhillon. H. S, Singla. R, Rekhi. N. S and Jha. R.(2009), "EOG and EMG based virtual keyboard: A brain–computer interface," in *Proc. 2nd IEEE Int. Conf. Comput. Sci. Inf. Technol.*, pp. 259–262.

[8] Hiley. J. B, Redekopp. A. H, and Fazel-Rezai. R. (2006), "A low cost human computer interface based on eye tracking," in *Proc. IEEE Conf. Eng. Med. Biol. Soc.*, vol. 1, pp. 3226–3229.

[9] Hori. J, Sakano. K and Saitoh. Y (2004), "Development of communication supporting device controlled by eye movements and voluntary eye blink," 26th Annu. Int. Conf. IEEE Eng. Med. Biol. Soc., San Francisco, USA, pp. 4302–4305.

[10] Johns. M. W, Tucker. A, Chapman. J. R, Crowley. E. K, and Michael. N (2007), "Monitoring eye and eyelid movements by infrared reflectance oculography to measure drowsiness in drivers," *Somnologie Schlafforschung Schlafmedizin*, vol. 11, pp. 234–242.

[11] Schmitt. K. U, Muser. M. H, Lanz. C, Walz. F, and Schwarz. U (2007), "Comparing eye movements recorded by search coil and infrared eye tracking," *J. Clin. Monitor. Comput.*, vol. 21, pp. 49–53.

[12] Shang-Lin Wu, *Lun-De Liao, Shao-Wei Lu, Wei-Ling Jiang, Shi-An Chen* and Chin-Teng Lin(2013), " Controlling a Human–Computer Interface System With a Novel Classification Method that Uses Electrooculography Signals", IEEE transactions on biomedical engineering, vol. 60, no. 8, pp. 2133 – 2141.

[13] Shen. M. W, Feng. C. Z, and Su. H. (2003), "Spatial and temporal characteristic of eye movement in human–computer interface design," *SpaceMed.Med. Eng. (Beijing)*, vol. 16, pp. 304–306.

[14] Sprenger. A, Neppert. B, Koster. S, Gais. S, Kompf. D, Helmchen. C, and Kimmig. H (2008), "Long-term eye movement recordings with a sclera search coil eyelid protection device allows new applications," *J. Neurosci. Methods*, vol.170, pp. 305–309.

[15] Watcharin Tangsuksant, Chittaphon Aekmunkhongpaisal, Patthiya Cambua, Theekapun Charoenpong, Theerasak Chanwimalueang(2012), "Directional Eye Movement Detection System for Virtual Keyboard Controller", Biomedical Engineering International Conference.

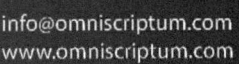

Printed by Books on Demand GmbH, Norderstedt / Germany